# Passionate About Prevention
## New Insights into Old Problems

By Tom Bissonette

With chapters by...

Bruce Newport
Jamie Utt
Robert Fuller
Sarah Newton

Passionate About Prevention - New Insights into Old Problems

Copyright © 2015 by Tom Bissonette

ISBN 978-1-329-74897-2

# Acknowledgments

Many thanks to Robert W. Fuller, Bruce Newport, Sarah Newton, and Jamie Utt for adding their insightful essays to this book. Thanks to Jane Bissonette and Hilary Evans for proof-reading and providing encouragement.

Finally, thanks to my four adult children and my students at The University of Tennessee at Chattanooga, since they provide the inspiration that makes me want to keep helping youth. They offer me a glimpse into a bright future by occasionally displaying their remarkable youthful wisdom.

*Tom Bissonette*

# Foreword

Prevention is an often misunderstood word. In common language it means to stop something from happening. In health related matters (including mental and behavioral health) it usually means to eliminate or reduce the conditions that make something more likely to happen.

This book mostly highlights primary prevention. It's much more ambitious to try to change the environmental factors that facilitate undesired behaviors than to intervene when something goes wrong. When it comes to prevention we frequently cannot muster the will or the resources to accomplish our goals. We are reactive instead of proactive.

This book contains a variety of ideas from several thinkers in the prevention field. It starts off with a history of societies' efforts to prevent alcohol abuse and an essay about bullying. Other chapters examine the issues of "rankism" and human dignity, a developmental perspective on harmful behaviors, and how to talk to youth in general. It appears to be a collection of random thoughts, but you will find a common and coherent theme that shows we need to look much deeper into these issues and understand the complexities of human behavior if we are serious about prevention.

Mostly this book is designed to trigger conversations in colleges and high schools about ways to make our prevention efforts produce better results. Prevention of harmful behaviors that affect youth is a daunting task. This endeavor is complicated by the fact that we want to encourage them to be independent, but also keep them safe.

From the 1960's forward colleges have attempted to respect children's rights and meet youth demands for more freedom. Higher education responded by moving from "in loco parentis" (essentially meaning substitute parent) to sink or swim. Students were given co-ed dorms; curfews were abandoned; less supervision was provided, and the party was on.

The responses in High Schools have been sporadic and often are influenced by local politics and wide disagreement on strategies, or even the basic philosophy of prevention. (e.g. Sexual abstinence only vs. comprehensive sex education). In many cases high schools have become more restrictive and less likely to encourage autonomy. Students making the transition from high school to college are often under-prepared for the extreme changes.

While the idea of granting more freedom is a positive concept, the problem is that it's based on the assumption that all students are equally ready for it. A main focus of this book is to address the fact that students vary in their degree of readiness and to explore ways to get those who are not ready more in sync with their life circumstances. Hence, most of the chapters will examine these issues from developmental or values perspectives and the internal changes that need to occur toward this end.

We need to enhance our prevention efforts, avoiding oversimplified notions such as the ideas that there are good behaviors and bad behaviors, good people and bad people, and if we can target these undesired behaviors or actors everything will be fine. We need to look beyond the behavior and deeply into the psyches of students to determine the positive intent of the harmful behaviors, help students acknowledge the underlying needs, and find better ways to get them met. In short, given the freedoms they enjoy, the best prevention strategies for today's students involve helping them become more self-aware and more self-directed.

You will be hearing from several authors on different issues, but it's helpful to keep some basic prevention terminology in mind as you evaluate each essay.

First, there are two main approaches to prevention:

1) Cocooning – This involves protective measures achieved by manipulating the environment and implementing safeguards such as more oversight or keeping negative influences away. Presumably, the more vulnerable the youth, the more protection they need.

2) Pre-arming – This involves less direct supervision and focuses on preparing students for whatever challenges we believe they will face. This is usually accomplished through education or training individuals to anticipate and respond to environmental threats.

Although most efforts at the college level lean towards pre-arming, some cocooning still occurs. When schools conduct orientation sessions for new students they indoctrinate them about rules and warn them about safety issues. An example of cocooning would be, "Don't go to this area of town because it's dangerous."

Pre-arming a student would sound more like, "There are areas of town that have high crime rates so use good judgement if you go there and we suggest ......."

Second, there are three levels of prevention:

1) Primary prevention aims to prevent problems before they ever occur. This is done by trying to eliminate the hazards that cause a problem. This could include attitudes and beliefs as well as more tangible things.

2) Secondary prevention aims to reduce the impact of problems that have already occurred. This is done by detecting and addressing problems as soon as possible to halt or slow their progress.

3) Tertiary prevention aims to soften the impact of ongoing problems that have lasting effects.

Using the above terminology as a backdrop for reading the rest of the book will prove helpful in organizing your thoughts and coming up with your own strategies and plans. If you are reading this, you're already interested in prevention, but I hope you are able to gain new knowledge and become even more "passionate about prevention."

Note: The final chapter is about my personal journey through education. Although it's not directly about prevention of alcohol misuse, drug misuse, or sexual misconduct; it does speak to the kind of caring environment necessary to achieve real and lasting prevention. It contains some thought-provoking ideas for students and teachers. Enjoy!

*Tom Bissonette*

# Table of Contents

# History or Hysteria?
## An Overview of Prevention Efforts

## By Bruce Newport

"Drinking without being thirsty and making love at any time, Madame, are the only things that distinguish us from other animals."
*Beaumarchis*

The quote above may be a bit overstated but it does point to a thorny problem. Since humans can make decisions, they can make bad ones. From the dawn of civilization societies have tried to regulate human behavior, yet, when and where did organized prevention efforts start and how have they developed throughout history?

That question is easier asked than answered. Prevention, as it relates to alcohol, drug, and behavioral health issues, is a relatively new term that wasn't used on a widespread basis until the late 1980's. Nevertheless, prevention related work existed long before that time.

Sometime between 971 B.C. and 941 B.C., King Solomon took the initiative to write the book of Proverbs which was later included in the Old Testament to provide wisdom for future generations. What set this writing apart from other biblical commands was the fact that this was written in the form of practical advice rather than a strict directive.

Nearly a thousand years after the creation of the book of Proverbs, the Apostle Paul addressed Christian followers from the city of Ephesus by stating, "Do not get drunk on wine, which leads to debauchery. Instead, be filled with the Spirit…" (Ephesians 5:18, NIV). Solomon, in his infinite wisdom, penned several insightful tidbits of advice regarding infidelity and drunkenness hundreds of years earlier.

About 600 years later, the Quran was penned and, within, warned followers about the evils of alcohol and gambling by declaring, "They ask thee concerning wine and gambling, say: "In them is great sin, and some profit, for men; but the sin is greater than the profit..."(2:219)

Moving forward in time from the mid 660's, things remained somewhat silent when it came to prevention oriented messages for almost 1,200 years. During these twelve centuries, most of us have become acquainted with stories regarding over-indulging kings, ale drinking Vikings, partying pirates, drunken military officers, and rowdy cowboys. However, it is extremely difficult, if not impossible, to find any historically documented examples of prevention related activity during this timeframe.

Starting in the early 1800's pastors of various denominations literally raced across a fledgling United States to convert as many people as possible. In those days, taverns and houses of ill repute were aplenty. As these pastors hopped from settlement to settlement, much of their messages contained warnings about the evils of alcohol and promiscuity. The effectiveness of these messages remains uncertain due to poor documentation. We can dig up lists of converts, but nothing was produced to evaluate the successfulness of their messages. Regardless, they helped plant the seeds of prevention by promoting abstinence.

Finally, in the early 1900's American's as a whole began to show an interest in addressing chronic alcohol and drug use. Not knowing where else to turn, U.S. citizens began to voice their concerns about alcohol and drug issues to lawmakers. Elected officials did the only thing they knew to do. They created taxation of certain products and made others illegal.

In 1906 the Pure Food and Drug Act targeted harmful and toxic drugs, and was later amended to include misleading drug labels in 1912. The latter was adopted, in part, to curtail the work of snake oil salesmen who traveled from town to town with magic elixirs that could cure everything from toenail fungus to heart disease.

The Harrison Tax Act of 1914 followed, which restricted the sale of heroin and was used to restrict the sale of cocaine. The legislation was meaningful, but had two notable flaws. First there was not any type of widespread campaign to educate the general public regarding the negative consequences of these drugs. As a matter of fact, it would be safe to say that the majority of Americans that lived outside of urban areas never even heard of the drugs. Secondly, law enforcement at the time was very inadequate. Meaningful communication and interaction from one jurisdiction to another was a major problem due to limited phone accessibility and information reliability. It seems absurd based on today's standards, but law enforcement agencies in several areas were limited to one automobile. It's hard enough to enforce drug laws with limited manpower, but doing it with access to just one vehicle is difficult to fathom.

In 1920 the 18th Amendment to the Constitution was enacted and Prohibition was on the books. Protestant lawmakers on both sides of the aisle and the Anti-Saloon League led the charge. Again, their idea of practical prevention was to simply eliminate the problem rather than address the root causes of alcohol abuse. The results were tumultuous. Crime became widespread and mafia activities reached an all-time high and so did murder rates. Law enforcement was totally unprepared to deal with the ramifications of this new law. Black market sales of alcohol soared. Moonshining grew from a few local sales to white-lightning runs across large regions of the U.S. Movies like to Untouchables made it appear that branches of law enforcement had the situation under control, but in reality it was just the opposite. Things became so bad that the general public and lawmakers realized a change had to be made before things became radically uncontrollable.

The 18th Amendment was repealed in December of 1933 ending prohibition. In order to celebrate this occasion, Anheuser-Busch, Inc. sent a team of Clydesdale horses to deliver a case of Budweiser to the White House. Leave it to a brewery to use a seemingly innocent gesture as a major marketing gimmick. For the next 80 years, Clydesdale horses have been used in numerous commercials, ads, and bill-boards.

A few years after the end of prohibition, perhaps one of worst prevention related films ever released was created and financed by a large church to help parents address the negative consequences of marijuana use to their teen children. It was originally released under the name, Tell Your Children, with ambitious intentions. However, it bombed and was shelved. Unfortunately, it was later discovered in a dusty corner of a library's archives sometime in the late 60's and purchased by a shady investor with pro-drug ambitions. His intention was to market the movie as a spoof. It was retitled Reefer Madness and the rest is history.

In the film, teens are shown using marijuana and turning into sex crazed zombies, committing crimes resulting in manslaughter, rape, suicide, and mental madness. The movie was so poorly produced that calling it bad would actually be considered a compliment.

After a brief period of mass marketing to 16-29 year-olds, it became a comedic, cult hit among high school and college students in the 70's and 80's. Students would enthusiastically throw "Reefer Madness" parties with the main event being the movie, during which inordinate amounts of pot was smoked.

The 1940's didn't demonstrate any notable prevention related activities within the continental United States due to our attention focused on World War II. On the flip side, tobacco companies wasted no time in taking advantage of the situation by massively increasing marketing efforts in the U.S. to broaden the use of their addictive products to stressed-out family members and spouses of those serving in the military. To complement their profit making efforts at home, cigarette companies such as Lucky Strikes played the patriotic card by providing small packages of smokes as part of food ration kits for our men in uniform serving overseas. What a great way to introduce young servicemen to cigarettes! It also kept current military smokers content until they could resume their full-fledged habits when they returned to the states.

Even though the general public didn't create any notable prevention messages during this time, there were a few notable prevention-related public service announcements and posters that were directed towards servicemen stationed in the U.S. and overseas. These marketing efforts almost exclusively outlined the dangers of STD's. One infamous poster is titled "Booby Trap" and features a buxom brunette flaunting her wares to two servicemen sitting at a table drinking a few brews. The bottom of the poster just reads, "Syphilis and Gonorrhea." I'm sure this poster was jokingly posted on a wall above a bunk in a military bar-racks.

Another poster geared towards servicemen features a pretty, well groomed, red-headed gal wearing a conservative outfit that reads at the top of the poster, "She May Look Clean, But…" It depicts miniature caricatures of servicemen from various branches looking at her. At the

bottom of the poster it reads, "Pick-Ups, Good Time Girls, (and) Prosti-
tutes Spread Syphilis and Gonorrhea." Just below that statement it reads,
"You Can't Beat the Axis if You Get VD."

Armed Forces Radio ran a few STD oriented ads, but they were relatively minimal. Remember, armed forces radio was designed to boost
morale. Talk of STD's was considered negative so these messages were
used sparingly.

Five years after the end of WWII, the U.S. engaged in combat during
the Korean War. Rather than create a whole new prevention campaign
for servicemen, some of the same posters and Armed Forces Radio announcements were rehashed and used during this time period.

The 1950's tended to be a little more aggressive regarding drug and
alcohol education. Unfortunately, the materials used during this time period would probably rate 3-stars on a scale of 1-10. Regardless of the quality of the posters and productions, credit needs to be given where credit
is due. This generation was the first to begin using new forms of media
such as radio and television. Even a few newspapers and magazines
jumped on the bandwagon, until they realized that running stories promoting alcohol, drugs, and promiscuous behavior was more profitable.

In the early 60's short film snippets began playing before featured
movies in selected theatres. The average runtime for these spots were
about thirty seconds to a minute. Almost every one of these ads showed
young boys approached by hoods that were wearing leather jackets with
their collars pulled up and hair slicked back. The hoods would offer a
marijuana "cigarette" or booze to unsuspecting kids and the unwilling kids
would refuse and then run away. A message would follow that essentially
reaffirmed that kids should not fall victim to peer pressure and find the

nearest teacher, policeman, or parent to report the incident. Of course in reality, the kids watching the movie clip knew the hoods wouldn't get caught and would most likely find the snitches. Of course, the greasers would then proceed to beat the honest kids to a pulp and make it clear that they would suffer a slow painful death if they ever snitched again. Hence the term, "Snitches get Stitches," soon became commonplace among the ruffians and is still used frequently today.

To complement the B rated movie ads, several anti-drug posters were distributed en masse during the fifties and 60's. Two popular posters stand out as being peculiar and somewhat confusing. One artistic work states, "If you smoke it, you will go insane." Behind the written text of this poster is a hypnotic, swirling vortex that is similar to something that would be seen in an Alfred Hitchcock movie. It is assumed they were referring to marijuana, but it never clarified what was smoked. Almost assuredly they were not referring to tobacco or corn stalks.

Another poster from the 50's and 60's shows a girl hiding in a wooden packing crate and reads, "Remember: kids that take drugs are losers." It is not clear why the young lady is hiding in a wooden packing crate. Was it due to shame or paranoia? Was the teen hiding from her parents? This question remains a mystery.

The late 60's and 70's may have been the most difficult time to promote any type of prevention oriented message. The troop strength of the Vietnam War had escalated from approximately 800 advisors in the early 60's to 550,000 when forces reached maximum capacity. Drugs were unbelievably accessible for the troops and punishment for their use was minimal.

As deaths exceeded 50,000 during the waning moments of the Vietnam War, the "Peace and Love" movement sprouted and then blossomed. With this movement came massive drug use and the development of new hallucinogenic and mind numbing drugs.

Alcohol was still consumed on a relatively heavy basis, but for many it was simply used to alleviate the uncomfortable side effects experienced while coming off some of the powerful drugs that were being introduced during this era. This time period saw modifications to existing drugs that made them more potent and popular. New modes of distribution made usage affordable and accessible. To make matters worse, drugs causing psychedelic trips and massive highs caused drug use to spike even further. Unfortunately, thousands of individuals never returned from their trips and highs. Deaths of musical superstars such as Jim Morrison, Jimi Hendrix, and Janis Joplin didn't curtail the growing popularity and obsession of these drugs.

So, how did society formulate prevention strategies to battle this new movement? One of the main steps was the creation of the "The War on Drugs" by government officials. It must be noted that movements such as this are not prevention in nature, but derive from desperation to stop supply-side strategies. This movement was, and is, very controversial.

Contrary to the belief of many, "The War on Drugs" didn't start with the Reagan administration. The movement actually started during the Nixon Administration. In June 1971, President Nixon declared a "War on Drugs." He massively increased the size and presence of federal drug control agencies, and put pressure on lawmakers to pass legislation such as mandatory sentencing and no-knock warrants. Although it was a temporary move, Nixon was instrumental in placing marijuana as a Schedule I narcotic, which is the most restrictive category of drugs.

In the 1970's campaigns and prevention messages went some-what dormant. When ads or materials were used to educate the public, it was just a matter of time before these campaigns were discounted by pro-drug advocacy groups and their associated propaganda. Any ads that did slip under the radar were shown as late night PSA's, which had minimal viewership.

Also during the 70's there was a major focus on racial equality and environmental protection. These were extremely important issues that, deservingly, garnered a great amount of public attention. However, the emphasis on both topics may have inadvertently lessened the focus on drug and alcohol prevention awareness for the time being.

In the 1980's, the Reagan Administration took Nixon's "War on Drugs" and massively enhanced and changed parts of the program. Massive manhunts utilizing the organizations such as DEA, CIA, FBI, and the Special Forces were tasked to capture and eliminate gangs, cartels, and laboratories that were manufacturing and distributing illegal drugs. The results were very mixed. Several key players were arrested and thousands of pounds of drugs destroyed. However, the flow of drugs into the U.S. and usage rate across the country didn't seem to slow down.

In 1984 along came First Lady Nancy Reagan with an idea. Rather that stop the flow of drugs, she wanted to focus on reducing the consumption of illegal substances through education and awareness. Programs similar to this had been tried, but not with help of influential people such as The First Lady.

First Lady Reagan's motto, "Just Say No" was ridiculed by the many in the press and the general public. However, to her credit, she helped stimulate a movement. Soon individuals and organizations across the nation began to recognize the need to rethink prevention oriented tactics as it relates drugs, alcohol, and self-destructive behavior. By the late 80's the public began thinking outside of the box and more on the demand side of the equation.

In 1987, a nonprofit, Partnership for a Drug Free America, aired a groundbreaking commercial. In the commercial an actor holds up an egg and says, "This is your brain." Then the actor cracks the egg and fries it in a pan. The frying pan is then held up to the camera and the actor chimes, "This is your brain on drugs...any questions?"

This commercial was considered brazen by many, but it broke through media barriers. The public was now faced with higher quality commercials that addressed drugs, alcohol, and other social ills.

In the early 90's people began to be exposed to anti-drug and anti-alcohol messages on billboards, radio, T.V., newspapers, and magazines. Some of the ads and PSA's were well received, while others fizzled. Regardless, Americans were now willing to accept that we had an actual problem and it was totally acceptable to publically talk about it.

Late in the 90's community groups, coalitions, and nonprofits began to emerge. Their common goal was to address drugs, alcohol, and other social ills. It was sometime during this time period that the term "Prevention" became a commonly used term.

Soon after, prevention groups began to grow exponentially across the United States. These groups were integrated into schools and universities, and government agencies began creating departments specifically dedicated to prevention. Even recovery oriented groups combined prevention strategies within their programmatic scope.

More than a decade after the prevention field gained substantial traction, the message of "Stop it before it starts" continues to gain favor. However, work in this area cannot stop here. As generations change, so do the challenges we face. In order to keep prevention on the cutting edge, we must continue to come up with new ideas and new techniques to counter the ever-changing problems related to drugs, alcohol, and behavioral health.

*******************************************

**Bruce Newport** *is a former Administrative Director for the Tennessee Certification Board, a prevention activist, and an avid history buff.*

# A Problem of Power: Ending Bullying in Schools

## By Jamie Utt

(This essay was originally published at Everyday Feminism)

People cannot stop talking about bullying. There are endless stories on repeat throughout the major media, and in the past few years, every state in the country has passed laws or policies that are aimed at tackling bullying.

Virtually every school in the country has a "Respect Week" or programming during October, National Bullying Prevention Month. And these conversations are important. They come from a deep and serious concern for our young people who are hurting. **But they are also grossly ill-conceived.**

Part of the trouble with tackling bullying is that there is no "one size fits all" approach, and there never can be one. And so long as we treat bullying as if it's some general problem that requires general solutions like "respect campaigns," we ensure that the problem of bullying will persist in our communities.

After all, at its root, bullying behavior is about power. Far too often, young people tear each other down and target one another for sustained violence, harassment, or neglect in order to feel more powerful, particularly when the person exhibiting bullying behavior is feeling powerless.

Amanda Levitt of Fat Body Politics describes it perfectly:

*If we actually started calling bullying what it is and address it as racism, sexism, homophobia, transphobia, ableism, fat phobia, and classism, it would actually give children a better way to deal with the very same power dynamics they will face as adults, while also giving adults more responsibility to challenge the intolerance that is rooted within our society overall.*

### Identity-Based Bullying

In essence, it's time we change how we talk about bullying. In my own work, I use the term Identity-Based Bullying to get at the root of the bullying problem.

Though there are, of course, exceptions, the majority of bullying in American schools cannot simply be explained away with "kids will be kids" or as "adolescent cruelty."

It is reflective of the very same problems of power, oppression, and privilege that we see in wider society, only it's played out in language and behavior that students can better understand.

After all, the patterns we see in bullying behavior reflect many of the issues of oppression and marginalization we see in wider society.

In Gender, Bullying, and Harassment, Elizabeth J. Meyer lays out the impacts of sexual harassment and body policing that young girls experience in school as one method of bullying.

The incredible researchers at GLSEN make it clear that LGBTQ+ students on the whole feel unsafe in school and are harassed and assaulted at alarming rates.

In their chapter "Fat Youth as Common Targets for Bullying" in The Fat Studies Reader, Jacqueline Weinstock and Michelle Kreibiel explain not only how common weight-based bullying actually is, but also how socially accepted it is within school climates.

In one school, students may be targeted for their race, in another for their physical or cognitive ability. In a third, they may be targeted for their religious expression or native language. Still in another, the bullying might relate to gender expression in subtler ways, with boys who are less athletic teased for their interests and girls who choose not to shave their legs tormented for their bodily expression.

The point, though, is that tackling bullying simply with "respect" and "kindness," while well-intentioned, simply misses the mark.

### Punitive Measures Don't Work

The most common outcome of the recent wave of anti-bullying legislation, though, has not been funding for trainings or curriculum that teaches students how to intervene when bullying is taking place around them or that gives teachers tools for building more inclusive classroom environments.

More than anything else, these laws hand down harsher consequences to punish bullies.

What these approaches fail to address, though, is that bullying cannot be solved with punitive consequences. First and foremost, punitive measures, though sometimes warranted, do nothing to prevent further bullying if for no other reason than pre-frontal lobe development in young brains.

If the part of the brain that helps us reason "If I take X action, Y will be my consequence" isn't fully functioning, then consequence-oriented policy isn't going to solve the problem of bullying.

Beyond simple biology, though, there are socio-emotional arguments to discourage "zero tolerance" punitive approaches to bullying.

Most students who exhibit bullying behavior are struggling and have been bullied themselves. In fact, among middle school students, the majority of students have participated in bullying behavior at some time.

Norris M. Haynes, Christine Emmons, and Michael Ben-Avie of the Yale University's Child Study Center even note that excessive punitive measures end up telling students who actually need more support that they are not wanted or welcome in the school community.

This is all to say that if we want to end the problem of bullying, we have to think differently about what solutions look like. In short, we have to transform the culture and climate of our schools by Building Cultures of Civility and Inclusion.

If we want to end the problem of bullying, we have to do two things: appeal to the rest of the adolescent brain, the part that relies on culture and habit; and address the specific nature of the bullying in our school environments by championing inclusion.

Educational researcher Sheri Bauman of the University of Arizona uses the term "climates of civility" to describe the challenge we face in tackling bullying.

If we want to end the problem, we cannot simply pass some laws and wash our hands. We have to do the tough work of changing culture and climate.

Fortunately, there are a few simple things that students, educators, and families can do to build cultures of civility and inclusion that prevent bullying.

## 1. Recognize That Every School Is Diverse

The first step to tackling the problem of bullying is acknowledging the diversity that exists in our schools. So often, the conversation about diversity is boiled down to simply race and class (with maybe some gender or sexual orientation discussed marginally). While these are vitally important aspects of student identity, they are simply part of the portrait of diversity in our communities.

Sometimes I will have schools in, say, rural South Dakota say to me, "We're not diverse, so we're not sure how the conversation about identity-based bullying applies to us."

**It leaves me baffled.**

Not diverse?

What about student ability? Citizenship experience? Weight and body image? Student interest? Religion? Gender expression? Sexual orientation? Race? Class and wealth?

The other side of the coin of comprehending bullying behavior is understanding the diversity that exists in each and every school. To paraphrase Gary Howard, "Diversity is not a choice. It's a demographic reality."

### 2. Treat Bullying as a Problem of Power

To tackle bullying is to tackle the specific nature of bullying in any given school community. To do that requires that we understand who is being targeted and what the bullying looks like.

More often than not, this is an exercise in understanding power. Students without social power are those far more likely to be targeted by others for bullying behavior, whether that's the social power of the school yard (i.e.: geeks vs. jocks) or the wider social power of identity privilege, power, and oppression.

When we understand who is being targeted, why they are more likely to be targeted in our specific community, and what this bullying looks like, we can begin to solve the problem.

### 3. Empathize

Empathy is vitally important.

We need to teach our young people how to empathize with others and how to stick up for one another, but we also need to model it.

Supporting those who have been targeted by bullying behavior is obvious (though sometimes it goes undone).

Far less popular, though, is empathy for those who've exhibited the bullying behavior. This is not to say that students shouldn't face consequences for their actions, but if we don't get to the bottom of why students are bullying, we won't solve the problem.

And more often than not, it's because a student is hurting.

### 4. Engage the Whole Community

Far too often, schools treat bullying as something "in-house." Parent engagement is an afterthought, and the "support staff" of custodial workers, office workers, or security staff is all but ignored.

Training students to be UPstanders instead of bystanders is rare, and teachers aren't often given the time to design school-wide interventions to tackle the problem.

Shifting culture and climate, though, means bringing everyone on board.

Offer families constructive ways to participate in the conversation. Take the time to train students and discuss bullying prevention in advisory. Offer all staff members opportunities to design and implement proactive and preventive solutions. Because as the old saying goes, "It takes a village."

### 5. Be Proactive, Not Reactive

So long as our approaches to bullying remain reactive, we will never mitigate the problem. We have to create the kinds of environments where students don't bully.

Doing so responds to the part of the student brain from which they are more likely working, the part that relies on environmental cues and habits, in building critical mass for change.

Nearly every aspect of student experience and achievement can be tied back to inclusiveness. When students feel safe and included in school, they show up for class. When students feel fully supported in their identity, they engage socially. When students are taught from an early age what it looks like to build inclusive environments, they are more likely to stand up for their peers. When students feel safe and included in school, they achieve at higher levels.

Simply put, we cannot punish bullying into oblivion.

We can, however, create environments where we value respect, empathy, care, and (at a minimum) civility. And when those things are valued, bullying simply isn't tolerated.

### Ain't No Easy Answers

When I'm approached by a principal, counselor, student, or parent to offer bullying prevention training or consulting, the client generally falls into one of two categories. Half want easy answers. They want a simple, ten-step solution to the problem. I can't help these folks much.

But the other half? They understand that bullying is complex and nuanced.

They understand that we cannot just lump people into categories of "bully, bystander, and victim."

They understand that punitive measures don't work.

These are the folks who understand the work it takes to shift culture and climate and are committed to that painstaking transformation. These are the folks who are most likely to realize powerful change in their community.

And this is the group that I hope you fall in.

                    ******************************

*Jamie Utt was born and raised in Western Colorado. Jamie has had a commitment to truth seeking and justice from an early age. He earned his Bachelor of Arts in Peace and Global Studies from Earlham College where he dedicated his studies to conflict resolution and Middle Eastern Peace Studies. While at Earlham, Jamie spearheaded and supported many social justice efforts and worked as a sexual assault survivor's advocate in a nationally-recognized sexual violence response, awareness, and prevention program. He is Founder and Director of Education at CivilSchools: Building Bullying-Free Culture.*

# The Poison - Rankism; The Antidote – Dignity

## By Robert W. Fuller

I got a close look at the poison of rankism at the age of seven, when my classmate Arlene was sent into the hall for the whole day. Arlene lived on a farm and wore the same dress to school every day. When she spoke, it was in a whisper. Our teacher, Miss Belcher, began every day with an inspection of our fingernails. One day she told Arlene to go to the hall and stay there until her fingernails were clean. I wondered how Arlene could clean her nails out there, without soap or water. If there was no remedy in the hall, then the reason for sending Arlene there must be to embarrass her and scare the rest of us. Later, filing out to the playground, we snuck glances at her. She must have heard the snickering as we passed—hiding her face against the wall, trying to make herself small. Arlene felt like a nobody. I told my mother what had happened to Arlene, and, as I must have hoped, she made sure it never happened to me.

Other kids whom my classmates regarded as safe targets for abuse included Frank, who was shamed with the F-word; Jimmy, who had Down's syndrome and was ridiculed with the R-word; and Tommie and Trudy who were teased about their weight. The N-word was used warily, typically from the safety of the bus carrying our all-white basketball team home in the wake of defeat to a school that fielded players who were black.

Not belonging to any of the groups targeted for abuse, I was spared—until I got to college. There I realized that higher education was less about the pursuit of truth than about establishing another pecking order. I found myself caught up in games of one-upmanship, and was reminded of my classmate Arlene once again.

The toxic relationships described above are all based on traits that mark people out for abuse, whether in terms of class (Arlene), sexuality (Frank), disability (Jimmy), body shape (Tommie and Trudy), color, or academic standing. Even with none of these traits you can still be treated as a nobody by someone who is simply trying to make themselves feel superior. It's called "rankism" and it's the cancer that's eating away at all our relationships. But there is no justification for rankism any more than for its manifestations in terms of racism, sexism, ageism, ableism, or homophobia.

Why Dignity Matters

Emily Dickinson spoke for Arlene and me, and for no-bodies everywhere, in her "nobody" poem:

I'm nobody! Who are you?
Are you nobody, too?
Then there's a pair of us—don't tell!
They'd banish us, you know.

As she notes, nobodies are on the lookout for allies, and are on constant guard against potential banishment. As social animals, banishment has long been tantamount to a death sentence. It's no wonder we're sensitive even to the slightest of indignities.

Dignity matters because it shields us from exclusion. It assures us that we belong, that there's a place for us, that we're not in danger of being ostracized or exiled. Dignity is the social counterpart of love.

Racism and sexism were long regarded as part of human nature, but it's now obvious that these forms of discrimination are losing their legitimacy. To be racist, sexist, or homo-phobic today isn't cool, it's embarrassing (not to say illegal). Is it possible that we could make rankism uncool too, or will "dignity for all" remain a dream?

In a seminal work of the modern women's movement, Betty Friedan wrote of "the problem without a name." A few years later the problem had indeed acquired a name – "sexism" and from then on women knew both what they were for (equal dignity and equal rights) and what they were against (indignity and inequality). That's why pinning a name on the behavior that poisons relationships is the first step towards delegitimizing it.

As president of Oberlin College in Ohio during the early 1970s, I saw a non-stop parade of nobodied groups find their voices and lay claim to equal dignity: African Americans, Asian Americans, Native Americans, women, homosexuals, and people with disabilities. In every case, the inferior social rank that had been assigned to these groups came to be seen as groundless. Our view of human nature doesn't change over-night, but it does evolve over generations. The process typically begins with martyrdom and culminates in legislation. In between come years of nitty-gritty organization. But once enough people stand up for their dignity it's not long until they become a force to be reckoned with.

The task confronting us today is to delegitimize rankist behaviors just as we are doing with other forms of oppression. That means all of us – you and me - giving up our claims to superiority. It means no more putting down of other individuals, groups or countries. It means affirming the dignity of others as if it were our own. Sounds familiar? It's the "golden rule" of dignity which rules out degrading anybody else. When denigrating behaviors are sanctioned, potential targets—and who isn't one at some point? —must devote their energy to protecting their dignity. A culture of indignity takes a toll on health, creativity and productivity, so organizations and societies that tolerate rankism handicap themselves.

Rankism—The Source of Prejudice and Discrimination.

The relationship between rankism and other "isms" based on race or sex is like cancer and its subspecies. For centuries the group of diseases that are now seen as varieties of cancer were regarded as distinct illnesses. No one realized that lung, breast, and other organ-specific cancers all had their origins in cellular malfunction.

Following this metaphor, racism, sexism, and homophobia are analogous to organ-specific cancers and rankism is the underlying malignancy analogous to cancer itself. Attacking each "ism" one by one is like developing a different cure for each kind of cancer, but to target rankism is to take aim at the root cause of prejudice and discrimination.

The cancer of rankism persists as a residue of our predatory past. But, for two reasons, the predatory strategy isn't working any more. First, the weak are not as weak as they used to be, so picking on them is less secure. Using weapons of mass disruption, the disenfranchised can bring modern life to a stop. Humiliation is more dangerous than plutonium.

Second, the power that "dignitarian" groups can marshal exceeds that of groups that are driven by fear and brute force. When everyone has a place that is respected, everyone can work for him- or herself as well as for the group. "Dignity for All" is a winning strategy because it facilitates closer cooperation. Recognition and dignity are not just nice things to have, they are a formula for group success, and their opposites are a recipe for infighting, dysfunctionality and failure. If we put the spotlight on rankism and purge our relationships of this poison, then not only we will spare people from humiliation, we'll also increase the creativity of our communities.

One of the sources of Lady Gaga's fandom is that she's a leader of the dignity movement. The kid who protests when one of his classmates is nobodied is another, all the more so if he or she is able to do so in a way that protects the dignity of the perpetrator. When victims of rankism respond in kind to their abusers, they're unwittingly perpetuating a cycle of indignity. The only way to end such vicious cycles is to respect the dignity of the perpetrators while leaving no doubt that their rankist behaviors are unacceptable.

In a dignitarian society, no one is taken for a nobody. Acting superior—putting others down—is regarded as pompous and self-aggrandizing. Rankism, in all its guises, is uncool.

Our age-old survival strategy of opportunistic predation has reached its sell-by date. A vital part of our defense against this strategy is not to give offense in the first place. Going forward, the only thing as important as how we treat the Earth is how we treat each other.

\*\*\*\*\*\*\*\*\*\*\*\*\*\*\*\*\*\*\*\*\*\*\*\*\*

*Robert Fuller is a physicist, a former president of Oberlin College, and a leader of the dignity movement to over-come rankism. He has consulted with Indira Gandhi and Jimmy Carter, and keynoted a Dignity conference hosted by the president of Bangladesh. Fuller's books on dignity and rankism—Somebodies and Nobodies and All Rise—have been published in India, Bangladesh, Korea, and China, featured in the New York Times, the Oprah Magazine.*

*He has just published The Rowan Tree: A Novel, a love story showing the spread of dignitarian values around the world. He has four children and lives in California.*

# How to Reach Young People

## By Sarah Newton

When people ask me how I manage to reach young people they think I possess some kind of magic pill or better still, a wand. Yes, I have been doing this for over 20 years. Yes, I am good at it and I hate to burst your bubble but it really is my desire to listen, hear and have a meaningful relationship with the young person that allows me to get through where others may have failed before. I would like to say that it is a Superpower and while I can often be seen wearing a cape, I believe that anyone can do what I do as long as they stop doing what they are doing first. I believe the reason that most coaches, parents, youth workers or teachers can't reach the young people in their lives is that they are too busy reading the books on how to do it or trying out a new worksheet they have found or even, heaven forbid, trying to use a "system" to get the young person to do what they want them to.

Reaching youth is often really coded as, "How can I get this young person to do what I want them to." Subtly, we try and control them and call it coaching, empowerment, support and a whole lot of other things. What if reaching young people just meant showing up, what if it was that radical and simple?

Well, I believe it is. I think that even in the pressurized and outcome-focused position that most youth professionals find themselves in, we can reach most of them by putting the papers, forms and techniques down and just showing up. I always remember the words a pas-tor once said to me regarding young people.

"All you need to do Sarah is love them more than they hate themselves." I thought that was the most beautiful and profound thing I have ever heard and still remember it often when I feel anything but love for the little angel in front of me.

I am trained in a style of coaching that is a little radical in its approach, focusing much more on the relationship then what you do with the client, if we can be firm and fair and have a meaningful relationship with our charge, we can probably support them to change their behavior in a much more powerful way. The work I do with young people focuses on a concept called The Quality World. In Reality Therapy, the quality world represents a person's total outlook and understanding of the world around them as it relates to people, possessions, beliefs, etc. Starting from birth and continuing throughout our lives, we place the people who are important to us, things we prize, and systems of belief (religion, cultural values, icons, etc.) within the frame-work of our Quality World. To reach a young person successfully we need to understand fully what is in their quality world, because every behavior we initiate will in our mind at least get us a real world experience of our quality world.

So, in easy to understand language, our quality world is how we see the world and is our filter by which we judge everything. Our quality world is what predicts how we will behave in certain circumstances and how we interpret events that go on outside us. Once we understand someone's quality world, it is so easy for us to support them in changing their behavior.

It seems so simple, yet it is often missed or glossed over. We work with students to increase grades, assuming they want them increased and often not even asking, we work with the unemployed to get jobs, not even asking if they want one, we assume a drug addict wants to give up and assume someone in a violent relationship wants to get out. We don't take the time to find out to ask enough questions.

The best example I have even heard of this is from an international development health worker who was going out to work with some youth to educate them on malaria. The people she was working with had mapped out how they were going to teach these children about the spread of this awful disease. Because she has worked with me previously on the Quality World concept, she insisted that before they do any training then got inside the children's quality world to find out what malaria meant to them. She went out and asked them questions and what she discovered was shocking. These children thought that malaria was spread by milk, not mosquitoes. They had to change the whole training in order to hit the message home. Had she not insisted in getting inside these children's quality worlds, their training would have been ineffective.

About 8 years ago, when I started working with a school to increase their top grades, we spent almost 6 months trying to figure out the Quality Mind of these students. They didn't see themselves as bright, they didn't think they could get A grades and getting into a good university wasn't even on their radar. All they wanted was C grades to get to the local college. If we wanted to change the Quality World of these young people we had to start from scratch, we had to help them believe they were A grades students and that they could go far. Teaching them how to get A grades at this point would have been useless as they hadn't placed A grades in their Quality World. Eight years later, the top students now have expectation of 6 or more top grades and getting into the best universities. We have changed the Quality World of these students by being consistent about the messages we put in there.

Imagine for a moment that we both have two large pieces of paper in front of us and a stack of magazines. We are told to fill the paper with pictures that meant something to us. Off we would go with scissors and the glue and come back with our masterpieces, which would be hugely different from each other.  Each one of these pictures would be a glimpse into our Quality World, a glimpse into what is important to us, the things we hold dear. On mine you would see films, tons of books, food and maybe even a little alcohol. You would see lots of color and open space. You wouldn't see me socializing much or doing things like rock climbing. You might judge my interest in films as shallow or my lack of others in my pictures as meaning that I am a loner and you might even view my cut-out picture of Baileys as an indication of gluttony. You would be wrong on all fronts. Yet we do this to young children all the time; we tell them they are wrong for gaming all the time, for hanging out with this person or that person, for being so interested in social networking, or being inspired by a rap artist. We make such judgements about the things that are important to them, the things they place in their quality worlds. If we really want to reach young people then this is where we need to start; we need to start with the things that they hold important.

How do I get through to young people where others might not? I link getting good grades to gaming. I ask a future YouTuber how he is going to make this happen, not why would he want to do this. I help a rape victim get over her ordeal by writing about it as I know she wants to help others. I help young people see the advantages to their ADHD, which they hold as an excuse so tightly. I become that influential adult in their quality worlds who thinks that education matters, who thinks that doing the right thing matters.

When you understand and appreciate a teens' quality world you can support change from that place, which is much more powerful than making an assumption that they want what you want.

A person's quality word will also tell us a lot about the needs that drive that person to behave and act in a certain way. In the work I am trained in, there are five basic needs.

Survival. I can assume that we have that covered. If a child is in survival mode, then doing any kind of empowerment work is almost useless until this is managed.

Love and belonging.

Self-empowerment

Fun.

Freedom.

Each person has a different level of each of these needs, a level that we can't change, only work with once we understand.

Most conflicts occur because someone's needs are not met and they are doing what they can to meet them. That is all we can ever do, behave to meet our needs and therefore fulfilling the pictures in our quality world. So a child with a high need for freedom will constantly break curfews, because a curfew feels like you are hemming them in. A child with a high need for love and belonging will always be trying to do what it takes to fit in.

When we start to work with young people with these needs, they can start to make more informed choices if they wish, as can those that surround them. For example, my child who has a low need for freedom asks for curfews and will always be home early. My child with the high need will push them, so they are always a bit looser with room for movement, as I know if I hem her in she will rebel.

Let me show you how it pans out with me and the relationships with my family.

I have a high need for empowerment, fun and freedom and a low need for love and belonging.

In the relationship with my husband we are well balanced; his need for empowerment is not as strong as mine so I can lead most things. His need for freedom is about medium which balances out my high and keeps us grounded. His need for fun is the same; his need for love and belonging is higher than mine and this could cause problems, which I have explained to him. He could easily decipher my lack of wanting cuddles and always being with him as not loving him, which is not true; it is just that this isn't as high for me. Both of my children have high needs for love and belonging, so they know to turn to my husband more for emotional comfort and if they want a hug from me it is best to ask as I don't really give them voluntarily. Both my daughters have a high need for fun, which means that we can really have a great time together and one of them is high in empowerment, the other low. As for freedom, the same; one high, one very low.

We could, left to our own devices, have many fights and rarely get along, but since the whole family under-stands this information it allows us to make sense more of what is happening. It allows us not to make another wrong for the way they are and understand that what they need is different to what we need.

Helping young people understand their needs and how they play out in their relationships and life empowers them to look at things different and perhaps make different decisions and even change their behavior.

So when asked how to reach young people, my response is to understand what is in their quality world and link all you do to that. Help them understand their needs and how they play out in their lives. And fundamentally become the one person who truly listens, who doesn't judge and who loves them more than they hate themselves.

*************************************

**Sarah Newton** *is known as an author, speaker, consultant, creator of "Teenology", and youth media expert form England. Sarah has shared her wisdom with millions who have tuned into her TV and Radio shows, followed her writing and listened to her thought-provoking talks. She has an ability to raise the expectations and aspirations of young people she comes into contact with and works tirelessly to support young people to plan futures that propel them forward. She can often be found wearing a superhero cape, talking about The Hunger Games, and using the metaphor of story to get her message across. She loves Heroes and Heroines, slays Dragons often, loves cakes and has a huge shoe collection.*

# Are We Failing the Test?
# Misguided Retention Efforts in Colleges

## By Tom Bissonette

"Some of us think holding on makes us strong;
but sometimes it is letting go."
*Hermann Hesse*

Sometime around the 80's or 90's the word "retention" started to replace the term "dropout prevention" in colleges and universities and has become a notable catchphrase in the past 30 years. The use of the term has increased, in part, because educational institutions have been experiencing shrinking revenue from federal, state, and private funding. To counter this, colleges and universities began to rely more on tuition and fees paid by students. In many cases financial concerns became near panic-level relating to potential attrition because funding became more directly proportionate to the number of students enrolled.

I take issue with the word "retention" itself because it focuses on the needs of the university to retain students for financial purposes and not always the long-term success of the student or campus safety. This tendency can also interfere with effective alcohol, drug, and behavioral health-related prevention. Would it be more sensible for an identified at-risk student to leave college if she poses a serious threat to self or others? Do some schools feel pressure to put so much effort into unsalvageable students that other, more reachable students might slip through the cracks?

Of course we want to help students whenever we can, but we need some kind of criteria to guide us regarding where to focus our energy and resources.

The word student "persistence" has sometimes been substituted for retention, but has not, itself persisted. Retention is not just an inappropriate word; it's also a misguided practice. It tends to be reactive; not proactive. To date retention practices have relied more on backend-intervention than front-end assessment and prevention. To remedy this and reduce attrition all students should be evaluated before and as they begin college regarding their overall readiness. This includes academic fitness but should also include physical, emotional, relational, and other facets of readiness.

Because of the short-sightedness of a retain-them-at-all-cost mindset, student discipline and prevention efforts have sometimes been swallowed up in the retention wave and often with undesirable results. It puts the staff in a tenuous position if a student's presence poses a serious threat because the staff is under marching orders to keep as many students enrolled as possible. Sometimes Resident Advisors or other staffers get trained (in effect) to look the other way after a few instances where they aren't supported in the discipline process.

I am not suggesting that all discipline cases should be handled the same way nor am I suggesting we reduce efforts to help all students persist. Further, those with special circumstances such as learning disabilities or mental illness should be given the best care possible.

I am saying, however, that attrition rates should not be a factor when we consider how and when to help students. I believe that assessing students early and matching them with the right resources is the best strategy.

This can be done in a non-intrusive manner and it can be integrated into the normal enrollment and orientation processes. Simply put, if you offer sufficient individualized support from the beginning, students will be less likely to leave voluntarily or be dismissed.

To illustrate how much financial pressure can influence disposition of cases consider the following true stories:

In the first instance a student was living in privatized student housing and had not passed a single class in three semesters. He was allowed to live in a campus apartment even though he was an active student. His academic failure only came to light after he brandished a knife in a threatening way towards other students. Since his parents were paying his tuition and rent on time there was no urgency in dealing with the situation until then.

In another case a student was in her last semester of college and her parents went bankrupt. She had paid four years of rent but was now two months late. She was abruptly evicted and spent the remainder of the semester sleeping on a friend's sofa or in her car. She managed to graduate under these harsh conditions but was left with a bitter feeling towards the university.

Colleges and universities need to get more serious about helping students persist for the right reasons. We already know why students leave, they either aren't ready for college or they're not happy there. It's time to stop over-analyzing the problem and get to work. This point was made at a meeting I attended while I was a College Counselor. There was a suggestion on the floor to do a two-year student study on attrition. At the time there were about seventy people in the room who had direct contact with students every day. I spoke up and said, "We have many people in this room who know most of the reasons students leave. We don't need another study; we just have to ask these staff." I was well aware that I was committing a heresy in higher education where studies are the solution for virtually every problem!

Attrition rates and revenue streams should not affect discipline. It's hard to be consistent when sanctions or services are unevenly distributed. Some students are more "entitled" than others. If you are a star athlete or your family is a major donor you will likely receive preferential treatment. The real irony is that the "privileged" student who receives the rescue is often harmed along with everyone else.

It's dwelling on the obvious to say we need to be concerned about people more than money, but it's not always obvious how to create and sustain a culture of care and assure a healthy, safe campus.

The horrible 2007 Virginia Tech tragedy may be a case in point. A young man with a history of mental illness and obvious signs of maladjustment killed 32 people and wounded 17 others. In the official Governor's report, the incident was largely attributed to inadequate mental health care and "not connecting the dots," in other words – poor communication between departments. The report under-emphasized the fact that an individual whose behavior was so bizarre that it frightened other students in the classroom was taken out of the classroom and given private tutoring. At the time I was struck with the absurdity of this but appreciated the positive motive to go to extraordinary lengths to help a troubled student.

My suspicions were aroused when I visited the Virginia Tech website a few days later and discovered documents about the institutions twelve major initiatives. All the documents were available except the document related to "Retention." Perhaps this was just mere coincidence but it left me troubled and wondering.

It also made me feel that we need to uncouple discipline and prevention from retention and fast!

Prevention hasn't been unsuccessful for lack of effort. There is a large body of work in the last few decades and some widely used ideas toward this end. Some new ideas will be offered later in this book but it is also useful to critique some of the current methods and practices that can could still be useful in addition to fresh approaches. Currently the most popular are social-norming and bystander intervention.

Social Norming focuses on the prevalence of healthy behavior rather than looking at the frequency and consequences of negative behavior. It identifies peers who are participating in a healthy lifestyle, which happens to often be the real norm on college and high school campuses. The information is publicized widely on campus to increase awareness of the positive data. It attempts to nullify the "everyone's doing it" excuse by providing this contrary proof.

Social-norming is primarily a universal technique. It focuses on a campus as a whole or a specific class such as a new freshman class. The University of Virginia in Charlottesville is one of the forerunners in using the technique. UVC started this program in the late 90's utilizing peer counseling, social events and information campaigns. Several other colleges followed suit and set up similar programs.

The overall results of this type of program have been somewhat unclear. This is primarily due to the inability to gather accurate data at individual schools, especially those with vast student bodies. The cost can be prohibitive at larger schools, while smaller schools may not have a large enough sampling pool to establish useful statistics.

What we do know is that college campuses in the U.S. drink at the same rate now that they did a decade ago. In 2001 the Harvard School of Public Health did a very comprehensive study that found that 44% of the nation's college students drank at "binge" or "heavy binge" levels during the previous year. Each year, while intoxicated, approximately 1,700 U.S. college students died on an annual basis due to alcohol related poisonings or accidents.

A little over 10 years later, the Center for Collegiate Mental Health gathered data on 100 college campuses and found that the number still sits at 44% percent for individuals engaging in regular binge drinking, defined as five or more drinks in a row for men, or four or more drinks in a row for women. Alarmingly, the number of annual college fatalities went from 1,700 to 1,900.

The other prevention technique, bystander intervention, has recently even caught the attention of president Obama's Task Force on Sexual Assault. This technique, which is actually decades old, has been repackaged and is increasingly being used separately or in tandem with social norming.

Bystander intervention is designed to prevent risky behaviors and sexual violence by encouraging an entire campus to take ownership of this problem and speak out when they witness potentially dangerous situations or sexist language, etc. Other benefits of this approach include the reduction of victim blaming and harassment. This type of program also sets the stage for a campus to create a new social environment.

Like social-norming, bystander intervention is frequently a universal prevention technique. It is often introduced to entire campuses as a catch-all but it sometimes targets specific groups such as athletes or fraternities. An emerging body of research seems to show some success, at least in increasing awareness.

The drawbacks of this technique are relatively obvious. It can be very intimidating to become an active bystander because of social pressures and, without a doubt, personal safety concerns. It is difficult to get someone involved if they fear being beaten or killed. Unfortunately, these are real possibilities.

Also bystander intervention relies heavily on external social pressure which is contrary to some developmental theory that makes the case that internalization of values is the most important aspect of youth development. Both Social Norming and Bystander Intervention are often met with a strong backlash effect. Nonetheless, these are important tools for a comprehensive program and we should continue to develop them.

Meanwhile we need to work on new ways to help students avoid harmful behaviors and stop working on the false assumption that retention efforts will magically keep our institutions safe for all students. For multiple reasons it's time to explore some radical changes in the campus environment. Let's start by college personnel accepting collective responsibility for fostering social competence, not just academic performance.

******************************

**Tom Bissonette** is the Executive Director of YoungAndWiser, Inc. and Adjunct Faculty at The University of Tennessee at Chattanooga.

*Tom is a Psychotherapist and a Developmental and Prevention Educator. He has taught courses on Adolescent and Young Adult Development at the University of Tennessee at Chattanooga for over 20 years. He is the author of <u>Sexual Civility – The Hot New College Romance</u>, a book on how to develop the capacity for real intimacy as a sexual being. During his 25 years as a psychotherapist and faculty he created an effective model for helping youth understand and navigate their own development.*

# Making Prevention an Inside Job

## By Tom Bissonette

"Liberty means responsibility. That is why most men dread it."
*George Bernard Shaw*

It's normal for young men and women to sometimes focus more on the desire for freedom than the need for responsibility. Youth also have an inherent need for at least a sense of autonomy even if they aren't ready for the real thing.

I'm reminded of a girl I knew in high school. She came from a very wealthy family (compared to mine) and she was caught shoplifting. This was perplexing to me at the time because I knew her parents would give her everything she ever wanted. So I posed the obvious question, "Why?" Her answer made perfect sense. She replied, "It's the first time I ever felt I accomplished something on my own."

Her behavior wasn't based on a desire to do something wrong; it was rooted in a need to do something to meet a legitimate need. We must respect this kind of motive if we want to help youth make good decisions. With the exception of the psychopath, there is some positive intent in everything we do. That's why I am promoting new ideas and terminology for prevention. We can't effectively help others by discouraging certain behaviors without providing an alternate path for getting needs met.

"Circumvention" seems to be a good fit for a more comprehensive approach. One meaning of this word is "to anticipate and counter somebody else's plans or reactions." Designing prevention strategies or programs without anticipating these predictable counter-reactions virtually assures ineffectiveness. Students will always find ways to get needs met despite our best efforts. When they understand their own motivations they are more likely to admit to and own these needs and find better ways to meet them.

Nonetheless, some will fight hard to maintain the status quo. When you declare war on something it fights back. Consider our societies' war on drugs, the war on poverty, etc. The harder we fight the more menacing they seem to become.

What these campaigns have in common is their shortsighted goal of changing human behavior from the outside to address immediate concerns, while neglecting to foster more self-directed transformation in the long run. All such efforts have a shelf life but tend to continue beyond the expiration date. Those in higher education who think they can still program and proselytize youth into behavioral compliance really don't understand adolescent and young adult development and they underestimate the creativity of youth.

Case in point - the most accepted and implemented strategy to mitigate some of the effects of excessive alcohol use is the idea of the "designated driver." In theory it assures that the one driving the car is sober, thus rendering others safer.

Initially campus groups embrace this idea and implement it forthwith. As time goes on the desire to 'party" unfettered kicks in (because we haven't sufficiently addressed the cultural and developmental issues) and modifications are made. As one fraternity officer told me, "At first we rotated the duty among the officers, but eventually we pushed the responsibility all the way down to pledges." What could possibly go wrong when the least experienced driver with the least amount of authority in the group is chauffeuring a car full of intoxicated, rowdy frat brothers?

An agonizing example of the limitations of the DD strategy occurred in Detroit, Michigan in 1998. After an outing just a few days after winning the Stanley Cup, two of the Redwing's star hockey players were serious-ly injured in a limousine accident. One of them would never play again. They had hired a driver so they could safely celebrate their victory. It turned out the hired driver had a drinking problem.

When I lecture and train students about this I often say, "If you were really serious about safety you would have a designated drinker; one person who gets intoxicated and the rest of the group could protect them." Of course it's said in jest but at first they often aren't sure if I'm serious or not.

Prevention efforts such as the Designated Driver are limited because they rely too heavily on social control. While this focus on peer influence is an important aspect of prevention it must be accompanied by more emphasis on positive personal development.

The same is true regarding sexual assault prevention. With the passage of the Violence Against Women Act and the President's Sexual Assault Task Force mandates - while sexual misconduct prevention efforts on campuses are required by law - success itself cannot be mandated. That's why we need to expand our efforts so we comply with both the letter and spirit of these calls to improve what we do. We also need to modify our approach and offer incentives to change behavior that are more in sync with today's youth.

Of course, even years before mandates, many efforts have been made on campuses to address these problems but the results are mixed at best. Using scare tactics or guilt-trips to discourage these undesired behaviors doesn't work. Encouraging students to serve as peer monitors or interveners has probably helped; but these approaches produce a backlash effect. Further, this decades-old "bystander intervention" model may be less suitable for today's youth because some researchers make the case they are "more egocentric" than previous generations. If these researchers are correct we must appeal to their self-interest to get their attention.

Demonstrating the connection between social and individual responsibility and long-term happiness will appeal to their wish to make their lives more enjoyable, yet we must make the case in a way that resonates for them to keep them set on goals that require postponing gratification in the age of the quick fix. Circumvention then, contains both social elements and individualized developmental elements that encourage social compliance, but also entices students to be self-aware enough to become willing agents of their own personal growth.

In order to achieve a more developmentally-based approach we need to address the paralyzing orthodoxy in the prevention field. These behavioral and social problems have often been narrowly defined as "misconduct" or "criminal" behavior. While technically correct, these view-points are over-simplistic because they assume intentionality and they tend to divide students into categories of "good students" and "bad students." While there are some students who are sociopathic just as in any population, many at the margins are coming of age without adequate knowledge or support. They are acting out without malice of forethought because they lack a clear roadmap and the gear to navigate the steep, winding path that ascends to relational competence and fulfillment.

To some prevention staff the remedy for problem behaviors is to make more rules and enforce them more consistently. Others believe the answer is to inform students about the down-side while ignoring the short-term benefits of these behaviors and why they persist. Yet others want to deploy armies of do-gooders to stop the scourge of bad behavior by blocking and shaming their misguided peers. Current models of prevention may rely too much on outdated ideas about contemporary youth who may be more motivated by self-interest and prone to short-term thinking than previous generations. Yet, even this well-documented pattern cannot inform our efforts because there are so many exceptions.

Along with the problem of defining a common profile of adolescent and young adult tendencies we also are dealing with very different individual personalities. Rules and policies, though necessary, will not stop the risk-takers - the rule benders and breakers who only feed on the exciting, high-stakes seduction of prohibitions. Others, while not generally oppositional, are developmentally and relationally challenged and are trying to reach social and sexual milestones without the skills to in-crease their status and acceptance in the peer culture. They are reliant on others to set their compass. They need to be able to chart their own course and it's our job to help them to do it. Additionally, engaging students to combat other students ultimately serves to set them apart and creates a serious rift in a campus community.

Many experienced Preventionists will tell you that there are three groups on any campus, each with different needs:

1)  Those who will offend no matter what we do because they are immersed in a lifestyle or are addicted/dependent. Some of these are intransigent criminal types but many may be reachable/treatable. (Earlier referred to as "indicated").

2)  Those who won't offend because they are developmentally ahead of some of their peers or are just too constrained by fear or inhibitions.

3)  Those who might offend because they are developmentally challenged and/or are susceptible to negative peer culture, but would gladly opt out if they had an alternative. (Earlier referred to as "Selective").

The best way to enjoy quantitative and qualitative success is to reach all of these groups, but especially the third group; helping them learn about themselves and why they feel pressures that may lead to impulsive decisions. This group can go either way and this is where we can have the most direct preventive impact. They're struggling for self-regulation

Developmental theorists, including Lawrence Kohlberg and James Marcia, make the case that internalized values are more stable and persistent than externalized, borrowed values. In fact, Kohlberg considered our level of moral development to be directly proportional to how much we have accepted our beliefs as our own.

James Marcia's diagram of "Identity Statuses" illustrates the types of moral stability and the origin of values we might observe in youth based on their level of identity development:

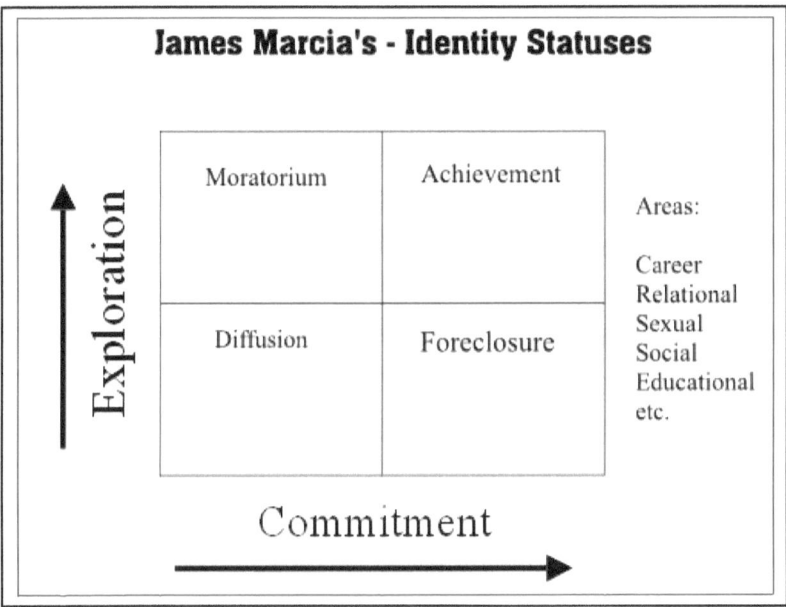

The Identity Statuses exhibited by specific youth matter as we consider ways to help them. We determine the status (type) by the amount of exploration done to arrive at an identity (vertical axis) and the degree of commitment to that identity (horizontal axis). Hence, the diagram shows that a person with little exploration and little commitment is "diffused" or confused about their identity. Those with high exploration and high commitment have "achieved" a clear personal identity.

"Moratorium" means one's identity is on hold while exploration continues. This may be the most common status for college students because they are constantly exposed to new information and encounter others from diverse backgrounds. It also represents a great opportunity for staff and faculty to influence students, but the competition is tough. We must understand that we are often trying to persuade them to give up something that they believe will give immediate pleasure or benefits for something that we believe might eventually pay off.

Those considered in "Foreclosure" have decided on an identity before they did sufficient exploration. They have identities borrowed from others, usually their parents. These identities hold as long as they are in environments that reflect similar beliefs and values. They are at risk when they are placed in environments that differ significantly from their comfort zone. The main concern is that if parents can assign them an identity then others can do the same. As I put it when I teach my classes about identity processes:

"If you don't have your own identity there are lots of people willing to give you one."

When we look at identity differences through the window of James Marcia's model it underscores the need for a variety of approaches. In order for prevention to work we must have sufficient information about individual students not just their peer culture. Thus we can help them nourish the side of them that wishes to make good decisions and suppress the side that leans toward self-defeating or short-sighted ones. Certainly peers affect this but a lot of the pressure is from within.

To the latter point, Lawrence Steinberg, known for his work related to understanding the "peer effect," has uncovered an interesting phenomenon. In an experiment to measure this effect he and his colleagues found no difference in behavior when peers were physically present to influence actions vs. when the subject believed that peers may be observing from another room. This strongly suggests that at least to some extent the peer effect is internal. If we accept this idea, then it's a game changer. It means we have to understand and help students understand what drives them in order to empower them to be the primary agents of their own psychosocial wellness. This can perhaps best be achieved through comprehensive developmental education.

# Youth Development: Can You Say Asynchrony

## By Tom Bissonette

"Man maintains his balance, poise,
and sense of security only as he is moving forward."
*Maxwell Maltz*

Most - if not all - of us in the prevention field appreciate the importance of understanding human development. Still, no matter how much we pay lip service to developmental theories and concepts, incorporating them into our prevention work is difficult.

In the last couple of decades many Student Affairs divisions at colleges committed themselves to changing their emphasis to "student development." Some even changed their names to include the word "development." They rebranded themselves to be centers where students could receive services commensurate with their needs at various checkpoints, usually based on grade level. They rightly concluded that the needs of freshmen are often different than sophomores, etc. Mostly these efforts fell flat, perhaps because development was too narrowly defined as academic and social competence.

We have been influenced by the great thinkers and researchers on the subject of human development, Freud, Erickson, Piaget, Kohlberg, to name a few superstars. They all have one thing in common; they see human development as unfolding in stages and define what is normal or typical for each stage. Positive results are determined by the level of conformity to these "normal" milestones.

This approach is sometimes referred to as "organismic" because of the "natural" stages and less emphasis on environmental influences compared to some other theories. While stages and milestones are helpful in comparing people to each other, they are limited when we try to examine the uniqueness and actual needs of individuals. If we adhere too strictly, we focus on what's normal instead of what's normal for this person at this time. We concern ourselves with the benchmarks that may tell us how far they need to go but not what they need to overcome to get there.

Understanding the asynchronous nature of human development is a step in shifting the focus onto the individual and understanding the pressures they feel and the resulting behaviors. This concept of "Asynchrony" means that people grow up at different rates. Some are ahead of their peers or classmates in some ways and may be behind in other ways. Of course some are about average.

Focusing on eight different developmental areas aids us to see how students or other youth view themselves and each other. These areas are:

Physical
Emotional
Cognitive
Academic
Social/Relational
Sexual
Moral/Ethical
Career

One could argue that other areas should be included, but these can be addressed separately. These eight main areas (domains) and how to assess them will be discussed and amplified later, but for now let's look at a few obvious examples of asynchrony.

Some students are more physically developed and mature earlier. Others mature later and look younger even though they are the same age. Another student may be way ahead of her peers academically, but may be behind socially. Since youth have a strong tendency to engage in "social comparison" (comparing themselves to others in their cohort), they sometimes feel pressure if they feel behind or inferior. Other times being too far ahead of peers creates problems. How many students "dummy down" and underperform academically to avoid standing out amongst their less gifted peers?

Asynchrony (or being out of balance) doesn't just apply when we compare ourselves to others, it can also mean we can be asynchronous within ourselves. Again, we can be ahead of ourselves in some ways and behind ourselves in other ways. This can be a factor in how youth put pressure on themselves to accelerate or slow down their progress. The mechanism for this is the "ego ideal" or, in other words, a different or better self we think we should be. This adds to the pressures and how we react to these pressures will determine whether we make good decisions or bad decisions.

So the ideas that human development is asynchronous, that we engage in social comparison, and we aspire to an ego ideal are the keys to self-understanding, which leads to making better choices.

To do effective prevention work or any intervention/counseling work we need to start where the client or student is, and that varies from individual to individual. I once met with a mother and daughter after the mother called and expressed concern about her daughter failing her classes. During the interview the daughter said at least three times she had been having thoughts about hurting herself. The mother ignored (or just wasn't ready to hear) the comments. At that point I asked the student to leave the room and I posed this question to the mother, "Don't you think it would be a good idea for your daughter to decide she wants to live before we concern ourselves with her GPA?"

This student was depressed but mostly due to her situation. Looking at this distressed student through a developmental lens, she was not emotionally ready for the stress of college even though she had the cognitive and academic ability. If we follow the logic of the asynchrony model we can assume there are other students who have similar profiles, and if they understood why they have difficulty coping, they could address this as a problem to be solved and skills to be learned as opposed to an external overwhelming force or personal failure.

Each student has some asynchrony and though it may not be as serious as the example above, it is possible to see each person as an individual. From a developmental perspective we can assess the issues(s) that each student is struggling with when we have a way to understand their unique profile. The Asynchrony Model doesn't diagnose pathology or judge people. We set the tone by telling students that asynchrony is normal and everyone shows it at some point in their lives. Practically speaking, since we cannot usually assess every student on our campus, we can show them how to do it themselves individually or in groups. This can be accomplished by asking them to assess themselves using the following diagram:

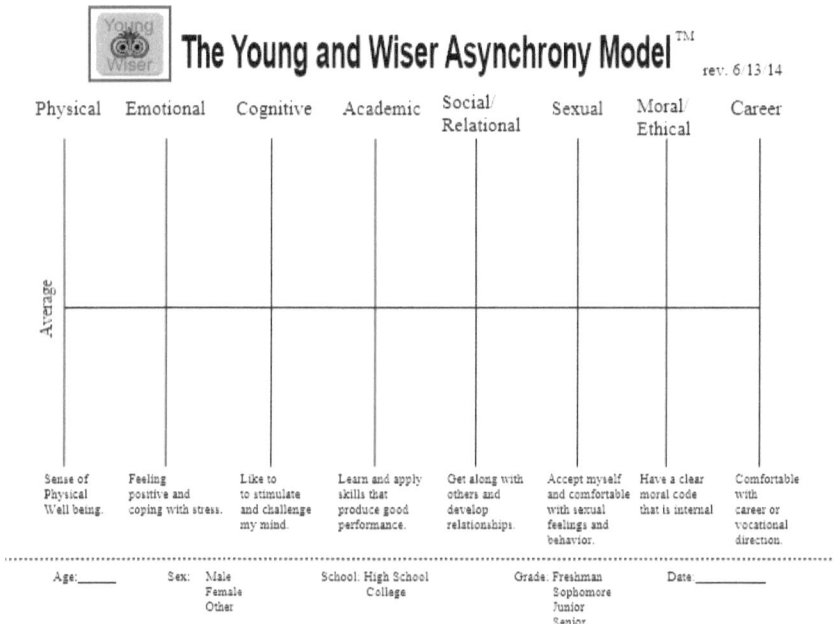

The diagram shows the eight domains of development and allows students to plot their perceived level on a continuum for each domain. We describe the domains to them by asking questions like:

Physical: Are you happy with the way your body performs and looks?

Emotional: Do you generally feel positive, and are you coping with stress adequately?

Cognitive: Do you like to stimulate and challenge your mind by solving complex problems, puzzles, etc.?

Academic: Do you find yourself spending enough time doing your school work, whether it is reading, studying or research, etc.? Do you have good skills in these areas?

Social: Are you making acquaintances and have some people you feel close to?

Sexual: Are you comfortable with your sexual beliefs and behavior? How well do they coincide?

Moral/Ethical: Do you have a clear moral code that reflects what you personally deeply believe?

Career: Are you satisfied with where you are in the career exploration process?

To keep it simple we ask them to compare themselves to what they believe is typical for their age instead of expecting them to be developmental experts. The average is defined as what they think is average or typical for their peers. It doesn't matter how accurate or objective they are at this point because we are interested in their view of themselves as they engage in social comparison. After all, it's their perception that determines the level of pressure they feel.

When they lack information about their asynchrony pattern, based on the level of pressure, they will either, 1) Step up: Develop appropriate goals to address the developmental asynchrony; 2) Catch up: Take short cuts that may lead to poor decisions; or 3) Give up: Feel hopeless and do nothing or act out. With information and increased self-awareness, as well as support, they can gravitate more often to Option # 1 above.

One student I worked with provides a good example of the first and most favorable response. This individual had severe deficits in social skills and extremely high levels of social anxiety. When we first met him he was literally unable to carry on a phone conversation long enough to order a pizza. After involvement in our social skills group for a year the growth was remarkable. That same individual was starting as a Graduate Assistant and was teaching a class at the university. How can such a profound change in such a short time be explained? With one word - Asynchrony!

His asynchrony was the problem and the solution. He was seriously delayed socially but very advanced cognitively. We simply capitalized on the fact that he was highly intelligent and he could understand how much his social issues interfered with his goal of becoming a biology teacher. All we had to do was provide a safe space for him to practice.

The level of pressures influences the degree to which a thoughtful, organized response will occur or a more impulsive reaction takes hold. It seems we need a certain level of pressure to be motivated but too much pressure can affect performance. The dilemma is how to increase pressure in some of the eight developmental areas and decrease it in others based on personal dreams and goals. Overall, we want to balance the pressure or keep it at optimal levels. This is sort of like the tires on a vehicle. If they are balanced, it's a smoother ride. If we increase the pressure a little, performance is better because of less resistance.

Alfred Adler said, "There is no such thing as talent. There is pressure." Of course aptitude makes a difference, but the argument about how much difference will never be settled and we're not going to try here. We do know, however, that those who achieve at a high level respond to pressure differently than those who don't. Sports fans know this from personal experience. Just recall the last time your favorite sports team was in a tight situation and the intense pressure you felt. Now imagine actually being in the game and feeling that ten-fold. Could you perform under that kind of pressure? Many athletes say they thrive on it. Some admit they can't deal with it and give up on sports despite being extremely gifted athletes. Perhaps for most of us the best answer is, "It depends on what the game is and how much we care about it."

Specific patterns of asynchrony create different but some-times predictable pressures. A student-athlete must emphasize fitness and skill development to excel in her sport. She usually has a demanding practice schedule. This takes time and energy away from other pursuits. She may suffer academically despite the special services she might receive as an athlete. For the sake of discussion let's stipulate that this particular student has difficulty or is bored with academic tasks. Because of her athletic gifts she has never really had to make academics a priority. Nonetheless, she has to perform at a certain level to remain eligible to play sports and be able to have a career which may or may not be related to athletics.

This student probably feels intense pressure related to her classroom performance. She may have solid morals and ethics generally but in this case she may be tempted to cheat. For her, lectures about policies and the ethics of higher education will not be enough. The problem must be addressed by either helping her find academic areas she is actually interested in or accommodating her learning style, or both.

Studies on cheating reveal that the power of the situation and opportunity are major factors in cheating. The peer effect also contributes. In a recent scandal at Harvard University almost half of an entire class was accused of cheating, involving 125 students. Seventy students were eventually suspended or expelled. This suggests that cheating is fairly common, even among our "best and brightest." Does this mean we should overlook cheating or lower or academic standards? No, we don't advocate that - but we do need to be more proactive with students who are vulnerable. This might include those on athletic scholarships, students who have to work too many hours to pay for college, and first generation students who don't have the family legacy of operating in academia. Also, all advisors need to be sensitive to the pressures their advisees feel. Thus, they need to understand the concept and assessment of asynchrony.

Let's look at another common pattern of asynchrony. In this case the student, Adam, has limited social skills and virtually no sexual experience. He is arriving at college - a highly sexualized setting - and hears lots of stories about the activities of others. He has heard many of the stories before in high school. He may feel intense pressure about being "inferior" or behind in his sexual and inter-personal development. Many coming-of-age, popular movies highlight this theme, often chronicling the adventures of youth as they try to "lose their virginity." There are often high levels of comparison and competition to say the least.

So our new college student begins to make some awkward attempts at meeting women and initiating sexual contact. After some failed attempts the frustration mounts. He looks around and sees other males who are "getting the women." They are not as nice or morally good as him and they don't even seem to actually care about the women they're involved with. You know, "Nice guys finish last", etc.

Somewhere along the way this "good kid" discovers alcohol and finds the solution to his problem. After a few beers he is bolder, more confident, and much less inhibited. Unfortunately, his judgment is now impaired as well. He could barely read social cues before but now we could say he's socially illiterate. During an "asynchrony" workshop I asked some students to predict what would happen. Before I could finish my question one spoke up and said, "Sexual aggression." This young man's profile is not uncommon on high school or college campuses and this is a problem we can and should anticipate. Further, this student is not the prototypical

criminal rapist we are determined to weed out. He would never hurt anyone sober but he may be dangerous when intoxicated. He's an appropriate
target for prevention/circumvention because he is in serious need of developmental education.

Now flip this scenario around and look at a victim. Karen. She has
been taught all her life that "good girls" don't initiate sex or let on that
they're interested. She has intense curiosity and has thought about putting
herself in the arena but really doesn't have a specific goal or plan. She
secretly (maybe even for her) hopes someone will take responsibility for
the decision so she won't have to. Obviously she doesn't want to be hurt
or humiliated but she is compelled to take a certain level of risk. She adds
to the potential serendipity of the experience by consuming a few drinks
herself. Along comes Adam. They begin a conversation and before long
they are laughing and exchange a kiss. Adam sees a green light and grabs
her breast. She has been assaulted! She leaves the scene and is traumatized.
It's not her fault but she may or may not feel that way. Adam doesn't get
it so he has another beer and mindlessly tries to find another opportunity.
He's on autopilot by now and he's headed for a rough landing.

Meanwhile Karen has yet another reason to fear men and her own
sexuality so she gravitates to this guy, David, she has talked with before.
David is strong and can protect her. She saw him punch a guy who was
mistreating a girl on another occasion. She doesn't realize that once he
accepts violence as a solution to problems then she is only safe until she
becomes the problem. Still she feels safe with him now and decides she
will let him be the first to have sex with her. Sometime later she ends up
in the emergency room after he beats her.

Is Karen to blame for her abuse? Absolutely not! A victim is never to
blame. Are Adam and David responsible for their actions? Yes, and no.

Yes, because in the eyes of the law they are guilty of crimes and should pay. If we allowed intoxication or common developmental delays to be offered as excuses, our entire system of justice would break down. On the other hand, Adam and David are victims of stifled development and cultural conditions. We do, after all, live in a rape/violence culture!

I repeat - we live in an insidious and pervasive rape/violence culture, yet we usually attempt to raise our young to behave appropriately and respect the boundaries of others. I recently underscored this point in a brainstorming session with students about how to prevent sexual assault when I gave an example of a typical dating ritual that constitutes sexual assault but is rarely treated as such:

"When in the presence of a potential sex partner it is common to touch them without asking permission. A young man fondles the breast of his date or acquaintance. She may push his hand away or tell him not to repeat the action, but technically, he has already sexually assaulted her."

Consensus on this observation was relatively easy to reach in this discussion group but one student said, "You're right, but how can we explain that to college students without sounding crazy?" He identified the main problem with normative behavior; it becomes invisible so pushing the cultural reset button seems extreme!

I saw a cartoon where a cave man was dragging a girl away from her parents' cave. He had knocked her out with his club. The parents were waving goodbye and the caption had them saying, **"Now you be sure to have her home by midnight!"**

(Think about the invisible norm in the-above cartoon caption).

(Then think about some visible norms today).

Invisible norms may include mixed messages from parents such as, "Violence is wrong but don't ever back down from a fight." In that case the invisible norm is that violence is OK in some circumstances.

Pertaining to sexuality we get constant invisible norms and mixed messages. Similar to the above-mentioned cartoon we get one very common message about sexual behavior. We hear that touching without consent is wrong but in common sexual "foreplay" we violate boundaries frequently and without thinking. (First base, second base, etc.).

If this type of ritualized behavior wasn't so common a modern dating scenario might go something like this: (excerpt from Sexual Civility – The Hot New College Romance)

Student #1 asks, "May I touch your breast?"
Student #2 replies, "One of them, or both?"
Student #1 – "Both, I guess."
Student #2 – "Do you mean touch or squeeze?"
Student #1 – "Both, I guess."
Student #2 – "How long were you planning to do this?"
Student #1 – "A few minutes I suppose."
Student #2 – "What do you want from this?"
Student #1 – "I just want to know what they feel like."
Student #2 – "Is this an experiment for Biology class?
Student #1 – "No, I just want to feel them!"
Student #2 "Let me see if I have this right. First you asked to touch my breast, then you wanted to touch both of them, then you wanted to squeeze them, then you wanted to do this all for a few minutes and it's not even for a biology experiment! The answer is no."

Of course the above dialog is absurd, but somewhere between absurdity and assault is a reasonable conversation that could establish boundaries without cultural scripts or games. In sexual encounters people should know the limits and boundaries ahead of time to avoid miscommunication and conflict. I often tell students that if they aren't comfortable talking about sex they aren't ready for it.

For young people benign curiosity can be a compelling force and it's not necessarily unhealthy. Curiosity in the context of a rape culture, however, becomes toxic and destructive. When a university fraternity (true story) sends out an email survey asking, "Who on campus would you like to rape?" we see curiosity and sexual fantasy infused with deep hostility. The outrage over this incident focused on a few young men behaving badly.

While this behavior needs to be directly addressed, the spotlight should also have been on whatever is happening in our society that makes them feel this level of hostility and insensitivity, and allows them to think - even for a moment - this is acceptable behavior. By the way, the campus where this occurred had a bystander intervention program that was considered so good it was copied by other universities.

Restated, the question could be, "What other cultural influences, besides ritualistic approaches to sex, contribute to the thoughtlessness associated with the rape mindset?" At the top of my list would be the invention and perpetuation of extreme gender norms and their worming their way into our collective psyche. To be clear, "Male and Female" are real biology-based), but "Man or Woman, Boy or Girl," are social constructs and roles to be performed. The once obvious advantages of these constructs were enjoyed because of efficient division of labor and not having to negotiate every interaction between males and females. No ten minute discussions about who should open the door.

I spent a few years solo in Ann Arbor, Michigan, where I socialized frequently and sometimes "dated." Because Ann Arbor had a high number of committed feminists per capita, I could never be sure if offering to pay for dinner would offend my companion or if not offering would offend. Once I became accustomed to this I was forced to have this conversation proactively and received the bonus of learning a lot about the person I was with very quickly.

Although it takes time and effort to have conversations, it's better than not having them and being frustrated and sometimes seriously harmed by miscommunication. The challenge is that when we choose to take this different, more open and direct approach it moves us outside the comfort of predictable gender roles.

So how does this mild discomfort turn into severe acting out? Males or females with high levels of insecurity about their role performance as a man or women need confirmation of this and will go to any length to get it. They are out to prove something! Thus they manipulate, harass, seduce, or rape their way into a temporary feeling of safe conformity. Our culture enables this by promoting the idea that males and females are so different they cannot possibly understand each other. Why bother with dialog if it will only lead to conflict or at least awkwardness?

If we are serious about preventing sexual misconduct, we must overcome these no longer useful biases. We will fail to come to terms with sexual mistreatment unless we recognize that it's a pervasive cultural problem, not just a few people or certain personality types misbehaving. We do a terrible disservice to ourselves and the casualties of abuse when we oversimplify the problem. Many people with sexual issues are victims of the hyper-sexualization of human relations and gender images that are too rigid. These behavioral and social problems have often been narrowly de-fined as "criminal misconduct." Despite the obvious truth in this, it nar-rows our choices too much when it comes to prevention or case disposition.

Because sexual transgressions are intertwined with common sexual behaviors youth are confused about how to establish and maintain inter-personal boundaries. Many prevention programs focus on shaming or scare tactics instead of positive relational training. Do we really believe that students don't know that sexual assault and rape are bad? Of course they know it's wrong, thy just don't know what it is!

Educators need more resources and new ideas. Victims need more options so they can seek resolution while avoiding the injustice of failed or re-victimizing criminal prosecutions. These ideas are not new; many have advocated for civil action, mediation, or so-called "restorative jus-tice" as alternate remedies.

On the prevention side, to some the answer is to make more rules and enforce them more consistently. Others believe the solution is to inform students about the down-sides of drinking and sexual aggression while ignoring the perceived short-term benefits of these behaviors and why they persist. Yet others want to deploy armies of do-gooders to stop the scourge of bad behavior by blocking and shaming their misguided peers. These strategies are unlikely to hit the mark because they trigger a severe backlash effect. Rules and policies, though necessary, will not stop the risk-takers - the rule benders and breakers who only feed on the exciting, high-stakes seduction of prohibitions. Additionally, engaging students to combat other students ultimately serves to set them apart and creates a serious rift in a campus community.

We can accomplish primary prevention by teaching developmental concepts in high school and college (or even earlier) so students can understand what drives their behavior and how to reach their personal sexual milestones without damaging others or themselves. We'll stop feeding the taboo-devouring fire or placing goodie-two-boots on the ground. No backlash effect will ensue because the battle will be internalized. Fostering cultural change in the both collective and individual psyche assures victory.

We can accomplish cultural change by discouraging sexual scripts and encouraging sexual dialog. We can teach youth that they can be sexual and civil at the same time. As mentors and guides we don't have to be sex-negative or sex-positive; we can be sex-neutral and relationship positive. We must drive home the point that sex takes place in the context of relationships, however fleeting or casual. Of course this is a challenge, especially to persuade young males to adopt this position since they are often socialized to see sex as an end in itself. On the flip side, many females do not take ownership of their sexual needs and are passive in sexual dynamics with partners. They are taught to feel shame. This was highlighted recently when I was teaching the Asynchronous Developmental Model to young adult women. When discussing their level of sexual activity one reported being very active while another stated she engaged in no activity. Interestingly they both apologized to the group! Shamed if you do and embarrassed if you don't?

So boys are taught to seek and girls are taught to hide. Most of us have heard the "men are hunters and women are gatherers" analogy. In human mating rituals it's more accurate to say men are "stalkers" and women are "trappers." I don't mean to say that all men are predators or women by-and-large deliberately try to ensnare men and trick them into relationships or marriage (despite many jokes we hear that paint these pictures). It's much subtler and less intentional than that. As with boys, "girl education" begins at home and in the school yard.

Like boys they have something inherently special and exciting in their sexuality but they have to pretend it doesn't exist on its own. For girls acting overtly sexual is taboo, yet they have to appear to have sexual potential, not actual interest. It's the opposite for boys who have to appear sexually interested and experienced, even if they don't have the actual history to strike such a pose. That's why girls feel they must look pretty or sexy whether or not they are interested in actual sexual behavior. (Incidentally, the word "pretty" in Old English meant "cunning" and "tricky"). Boys react to this 'trick with mirrors' by imagining they are the key to unlocking all this incredible suppressed sexual energy. A potent polarity indeed! So she has to act disinterested even if she is and he feels pressure to initiate sexual activity whether or not he has sufficient experience or desire to do so. This accident is waiting to happen.

"Sexual Civility" provides a new paradigm for navigating these complex coming-of-age interactions. It's based on some sound developmental principles. It requires a fresh look at how gender norming affects perception and behavior. Here is an excerpt from the book "Sexual Civility – The Hot New College Romance":

"Sexual Civility involves telling the truth when the truth needs to be told and choosing the right time and place. Direct, clear communication is vitally important when we are beginning to date or hang out with potential sex partners. What we are really doing at this point is negotiating boundaries. This process defines the character of the relationship we will have with this new acquaintance."

Sexual Civility requires advanced relational skills and a clear sexual identity. By these we mean the ability to communicate our wants and needs openly and honestly and the ability to set our own social and sexual agendas without performing an insincere role. Needless to say these qualities are not always present on the college scene or in teens and young adults generally.

A third type of asynchrony appears frequently on campuses. Often there is a gap between spiritual and social/relational development. The extreme example would be the street preacher who shouts at passers-bye but never really relates to them. He lumps them into a category and condemns them to hell. Very efficient, but it's probably not conducive to converting them to his way of thinking.

This phenomenon shows up in a subtler way for many students. They have come from strongly religious backgrounds and may feel uncomfortable with campus diversity and free-thinking. They may do well if they find enough like-minded individuals for support and affiliation, but they may otherwise be overwhelmed, isolated, vulnerable, or all-of-the-above.

I'm reminded of a friend who had strong religious ties and strong traditional family values. His daughter attended the school where he worked rather than move to a faraway campus. With her family and familiar social network, she would be safe. She ended up pregnant in her sophomore year and soon after married and left school. There are several possibilities that could explain this outcome but it's likely that she had strong social/relational and sexual needs that were outpacing the internalizing of her personal values. People with borrowed values are on borrowed time.

I've seen this in other ways with students. Parents would send them to private schools to avoid the "sins" of sex, drugs, and rock and roll, but on occasion students would confess to me that they just had better quality of all these things at private school.

Another tragic example of asynchrony occurred with a counseling client of mine who impregnated another student a decade before and married her even though he didn't want to. His religious upbringing was very incongruent with his social and sexual development. He explained it this way:

"In my family sex was dirty and certainly sex before marriage was seriously not ok. I wanted to have sex but I couldn't plan for it. Being prepared with protection would have meant I was planning to sin. In my family the only thing worse than sex was pre-meditated sex!"

Think of it this way. If different parts of us are highly asynchronous (in sexual development or any other area) reasonable self-compromise is difficult if not impossible. In essence we are at war with ourselves and we can't win without also losing. The Asynchrony Model is a giant step towards gaining balance and more mindfulness in our decision-making.

Although this chapter focused on just a few examples of asynchrony, there are countless application of the model. My hope is that someday all student behavior will be viewed through a developmental lens. I also dream of a time when high schools, colleges, and universities embed developmental life education in their curricula.

# My Educational Journey

## By Tom Bissonette

In the early 1950's I experienced my first year of school in a rural area in Northern Michigan, six miles out from the nearest town. Three of my siblings and I attended the same school, which was literally a building with one classroom, one teacher, and eight grades. And no - we did not have to walk for miles in the snow to attend school – we had skis, and even busses.

The teacher's surname was Fitzpatrick but we called her "Miss Fitz." She was at least in her late 60's at the time and she taught well into her 70's from what I heard later. Of course it would have been impossible for her to teach all these students simultaneously. Necessity forged the educational model and she did the only two things she could – delegate and inspire. She did them both very well. Every new student got the same message on their first day; that each grade had the responsibility to teach the grade below it, and above all we don't hurt anyone or let them get hurt.

There were unspoken rules too. One day I noticed some students sneaking out a window to go play in a cow pasture. With the moral absolutism of a four-year-old I felt duty bound to report this transgression to Miss Fitz. I was a bit perplexed when she replied, "I guess they decided they could learn more outside today." That statement became the basis of my absorbing her implied third principle of education, namely that a teacher should never interfere with a learning opportunity.

Later in the year, overtaken by my own restless craving for adventure, I exited through the same window of opportunity only this time it involved pigs and lots of mud. When I returned to the classroom I was confident I wouldn't be confronted about my absence but my wet, muddy clothes would surely elicit some kind of disapproval. Miss Fitz - never short on surprises - just gave me a hug and said, "I think there are some clothes in the closet that might fit you if you want to change." As I walked away I turned back and noticed I had left a sizable mud stain on her dress. I felt a little ashamed and wondered what she would say.

She never mentioned it...

By the time I was ready for first grade we had moved to a suburban area of a midsized city. The transition to a grade-based school was fairly smooth for the first two years. We sat at round tables in groups and had ample opportunity to move around the classroom; plus, there was playtime!

When I entered the third grade the structure was painfully different. We were shackled to individual desks and the teacher spoon fed us information whether we were able to swallow it or not. One day I just couldn't tolerate it so I took the eraser on my pencil and pressed on a tooth repeatedly until my gums started bleeding. I was allowed to go to the bathroom to rinse my mouth and I stayed there as long as I could. I vividly recall looking in the mirror, seeing the sacrificial blood, and feeling proud of myself for being so clever in finding a way to escape the torture of Ms. Chase.

Fourth grade was even worse! The desks were still there and the teaching was even more unidirectional and regimented. The pace of instruction was furious and I noticed that some other students were falling behind and I could see the intense frustration on their faces. One day a student asked a question and the teacher answered it but it was clear to me that the student was still confused. I got up from my desk to demonstrate how to do the math problem and the teacher quickly descended down the aisle and grabbed me by the arm and took me outside. She shook me violently and said, "You will never do that again!" I was terrified but much worse than that, I learned a lesson that day that no one should ever have to learn from a teacher. Her violence ran through me like high voltage and it lit up an area in my brain that had been dormant. For the first time I knew it was possible to feel hate.

Fifth through eighth grades were relatively uneventful if you don't count emerging puberty and all that goes with it. That is another story and, although related, it would take up too much space here. My classroom landscape of the period was sprinkled with moments of occasional inspired instruction sandwiched between episodes of paralyzing boredom. For multiple reasons my education up to and through then had not prepared me for the large high school in an adjacent city that was the only public high school available. You know the story - small fish getting swallowed up by bigger fish in the deep dark waters of teenage drama and cruelty.

My parents decided I would be better off in the small Catholic school just a few blocks from our house. I transferred there in my sophomore year and things improved socially and I could compete in sports in that "Class D" school. Nonetheless, my problem of boredom persisted and a couple of unspeakable tragedies that year sucker punched my whole school into profound numbness. President John F. Kennedy was assassinated and six of our faculty were killed in a head on collision within weeks of each other. We lost our hero and whatever sense of security we had left in a short span of time. So when I wasn't skipping school I would keep myself amused by debating with the teachers. Some took this better than others. My geometry teacher would praise me, especially when I could complete problems that even he couldn't master. Another faculty, a priest, didn't appreciate me debating theology with him and he suspended me for three days for arguing with him in class.

After that I decided to keep quiet and stay off the radar enough to stay out of trouble. I managed to keep the expectations of the faculty low. They didn't know of what I was capable, and I was more subdued so they left me alone. I had a good thing going; I could come and go as I pleased. Occasionally I would make the mistake of bringing attention to myself but not very often. For example, in one class I accidently got inspired by the challenge of writing a short story with the requirement of submitting it the very next day. I turned in my story only to be accused of plagiarism because no one with my low grades "could have possibly written it".

I made one other mistake in high school and that was allowing a friend to convince me to take the SAT in my senior year even though college was not solidly in my plans. When the scores arrived, I got a surprise visit from the Principal who came to the class where I was, literally pulled me up out of my desk by my ear and marched me to her office. I had gotten the highest verbal score and the second highest math score in the class. When she ordered me to sit in the chair across from her desk I thought to myself, "I would have probably gotten the highest math score too if I hadn't drunk so much beer the night before."

She proceeded to give me the classic lecture about how much "potential" I had and what a waste etc., etc., ad nauseam. My rather rude response was, "So I might be worth something one day but I'm not worth much right now?" To this day I'm relieved this nun didn't fly over the desk and strangle me. She left an impression, though, because of the anger she expressed. It was almost as highly charged as my fourth grade teacher's tirade, but this was different. I could tell she cared about me.

With my cavalier attitude about schoolwork, and a penchant for partying, my high school career was a wash. I graduated with a D average and virtually no study habits. The year I graduated the Viet Nam war was in full swing and the draft was on. I enrolled in a community college, but even the fear of mortal combat wasn't enough to override my complacency. I dropped out because I was failing anyway and decided to surrender my student deferment. Draftees were chosen by lottery and my number came up quickly. I blew through two years of military service, fortunately never saw combat first-hand, and ultimately came out of it with the wind of the G.I. Bill at my back.

I started college again if for no other reason than to collect the money. All I had to do was maintain a C average and that's exactly what I did. I had a job at the time that reimbursed me for tuition so I was making a profit from this venture by double-dipping. Somewhere in my third year I made another mistake - a big one! I had hung around long enough to contract the disease I call "scholaria." I had always been intellectually curious but had never been intellectually disciplined. By that I don't mean things like getting the work done on time, or jumping through hoops like a trained porpoise; I mean focused intellectual effort driven by a deep sense of purpose.

Ultimately the bargain I struck with myself was that I could keep going as long as I believed that my efforts would have a real effect on the world. Newly endowed with this passion I started to think about grad school. My earlier undergrad grades made it difficult to get accepted in a top-tier institution but that didn't stop me. By the time my circumstances allowed me to apply a few years later I had decided on the social work/counseling field. Since I lived in Ann Arbor near one of the highest ranked Social Work schools in the nation I applied there. I was flatly rejected!

I was determined enough to re-apply the following year with better references and a cover letter that grabbed their attention. In it I stated, "I hope you will accept me this time because I'm going to keep applying until you do." Apparently that display of hubris got me the interview and the rest was easy.

My grad school tenure was marked by academic and practical experiences of the highest quality. Just as important, it forced me to learn how to use computer technology, something not easy for everyone in my generation. Once I saw what computers could do I was certain they would transform the world and I had to get in stride or become irrelevant. Ever since then my unabashed love for technology has extended my career and my reach.

Continuing a long and enjoyable career of helping others, I still teach at my university and run a nonprofit organization that serves youth and helps them – you guessed it – "reach their full potential." Just like my best teachers and mentors we don't shame them or scare them, we simply tell them the truth and love them while trying to get them to hang around long enough for something to happen.

Given my lifelong journey as a student, counselor, and university faculty I have distilled my experience into a few principles that I believe should inform our practice as teachers at all levels of education. With proper credit given to Miss Fitz and others I submit the following prescriptions of pedagogy:

1)   Students should be given more responsibility for their learning and the learning of others. One way I promote this is by giving my students jobs at the beginning of each semester. They have titles like "local news hound", the student who reports on local events that are relevant to the subject, or even "the class clown" who must come to class each week with jokes. I tell the clown that the jokes better be relevant, but if not, then they must be exceptionally funny.

2)   No one gets hurt. This means physically of course but also emotionally. Our rules of engagement in my classes include a requirement to honor the intelligence of others. We arrive at conclusions based on what we've learned thus far. Bad ideas are just good ideas under development.

3)   Everyone learns from everyone. As a teacher you are only one source of information. With the proliferation of the internet you're probably not even the most up-to-date source.

4)   Students have no obligation to learn or get high grades. We can complain all we want about students not caring but they have the right to get what they want out of college. If they want to barely pass to get the degree to enter a field of their choice, that's fine. If they are in school just to keep parents off their backs, that's a good reason too.

5)   Stop fighting technology. It's just the irresistible cow pasture of today. But keep fighting to assure it is used for good purposes. For example, I allow texting in my class but give parameters. They may text if it relates to helping a peer understand the lecture or if they are experiencing an urgent matter at home. They are on their honor, and although they could misuse this opportunity, they have the right not to pay attention - especially since they already paid tuition. As long as they don't distract others, I don't concern myself with it.

6) Be collectivistic and individualistic in assignments and exams. Example: Besides doing group projects, my final exam is 50% based on voting on the answers and 50% based on individual effort.

The idea that all we ever need to know we learned in kindergarten has long been an American truism thanks to the popular 1980's book by Robert Fulghum. With this basic wisdom and a more active role in our personal development during and after childhood, we are assured of a more fulfilling and productive future. If we just add a few more grades, some life experience, and the lessons garnered from being a teacher or an active learner we might expand Fulghum's dictum just a little by saying, "All we ever need to know we learned in the one-room schoolhouse we call "our world." The academic part of that world should be collaborative, rich, diverse, and sometimes muddy; which is to say that in one way or another, we should all be Miss Fitz.